WORK

Rebecca Woodbury, Ph.D., M.Ed.

Gravitas Publications Inc.

WORK

Illustrations: Janet Moneymaker

Work
ISBN 978-1-950415-20-5

Published by Gravitas Publications Inc.
Imprint: Real Science-4-Kids
www.gravitaspublications.com
www.realscience4kids.com

RS4K

Photo credits: Cover and Title Page, By New Africa, AdobeStock; P.3. By nenetus, AdobeStock; P.5. By Monkey Business, AdobeStock

Have you ever heard your
mom or dad say,
"It is time to go to work"?

Do you do work?

I don't think so.

Have you ever heard
your teacher say,
"I'll help you with your work"?

Can you help me do math?

I can find cheese.

But what is work?

That is a
good question!

Work?

In **physics, work** happens when **force** • • • • • • • • • • ▶ moves an object.

I can use force and work!

Review: FORCE

Force is any action that changes:

- The **location** of an object,

- The **shape** of an object,

- **How fast or how slowly** an object is moving. (This is called the **speed** of an object.)

When you pull your sister in a wagon, you are using **force.**

Because you are moving your sister, you are also doing **work.**

I have lots of sisters!

What happens if you pull two sisters in the same wagon for the same distance?

You use **twice** the **force,**

AND

you do **twice** the **work.**

See!
I was right!

What happens if you pull
three sisters in the same
wagon for the same distance?

Three???

You use three times as much force?

That sounds right.

You use **three times** the **force,**

AND

you do **three times** the **work.**

I was right again!

What happens if you pull four sisters in the same wagon for the same distance?

How much **force** will you use?

How much **work** will you have to do?

I think you will use four times the force.

Yes! And you will do four times the work.

How to say science words

force (FAWRS)

location (loh-KAY-shun)

physicist (FIZ-uh-sist)

physics (FIZ-iks)

related (ree-LAY-tuhd)

shape (SHAYP)

speed (SPEED)

work (WERK)

www.ingramcontent.com/pod-product-compliance
Lightning Source LLC
Chambersburg PA
CBHW040153200326
41520CB00028B/7595